▪THE ▪ WAY ▪ IT ▪ WORKS ▪

Air

PHILIP SAUVAIN

HEINEMANN

© Heinemann Educational Books Ltd

Heinemann Educational Publishers
Halley Court, Jordan Hill, Oxford OX2 8EJ
a division of Reed Educational & Professional Publishing Ltd

MELBOURNE AUCKLAND FLORENCE PRAGUE
MADRID ATHENS SINGAPORE TOKYO
SÃO PAULO CHICAGO PORTSMOUTH (NH)
MEXICO IBADAN GABORONE JOHANNESBURG
KAMPALA NAIROBI

British Library Cataloguing in Publication Data
Sauvain, Philip *1933–*
    The way it works: Air
    1. Air
    I. Title
    551.51

ISBN 0-431-00748-9

Photographic credits
t = top b = bottom r = right l = left

4 ZEFA; 5 C.B. Picture Library; 6 B.U.P.A.; 11, 15 ZEFA; 17 Robert Harding Picture Library; 23 ZEFA; 27 Robert Fowler; 29 Japan Shipping Centre; 31, 32, 33 Science Photo Library; 35 ZEFA; 36 Quadrant Picture Library; 38 Science Photo Library; 41*t* Quadrant Picture Library; 41*b* ZEFA; 43*t* NASA/David Baker; 43*b* Science Photo Library

Designed and produced by Pardoe Blacker Limited, Lingfield, Surrey, England
Artwork by Terry Burton, Tony Gibbons, Jane Pickering, Sebastian Quigley, Craig Warwick and Brian Watson
Printed and bound in Spain by Mateu Cromo

96 97 98 99 10 9 8 7 6 5 4 3

Note to the reader

In this book there are some words in the text which are printed in **bold** type. This shows that the word is listed in the glossary on page 46. The glossary gives a brief explanation of words which may be new to you.

Contents

What is air?

There is air all around us, but we cannot see it. This is because air is made up of different **gases** which have no colour. The main gas in air is **nitrogen**. Air also contains a gas called **oxygen**. All living creatures need air. They use the oxygen in air to stay alive.

The layer of air around the Earth is many kilometres thick. We call this layer the Earth's **atmosphere**. The weight of the atmosphere presses down on everything on Earth. This is called **atmospheric pressure**.

Living together

Food, coal and oil are all fuels. We burn fuels to make **energy**. Fuels need oxygen in order to burn. We use the oxygen in the air to turn the food we eat into energy. All animals breathe in air to get oxygen. They breathe out a gas they do not need called **carbon dioxide**. Plants do need carbon dioxide. It helps them to grow. Plants take carbon dioxide from the air and give off oxygen. So plants produce the oxygen that animals need and animals produce the carbon dioxide which plants need. The plants depend on the animals and the animals depend on the plants. This is called **symbiosis**.

How air behaves

Although we cannot see air, we can feel it when it blows as a wind. The wind blows when cold air rushes to take the place of warmer air. Warm air rises and cold air sinks. You can feel this in a warm room on a cold night. The warm air rises and escapes through gaps in the tops of the doors and windows. This makes room for cold air to blow in under the doors making a draught.

We use the wind as a source of energy. The wind turns windmills and pushes against the sails of sailing ships. You can see the power of air when you blow up a balloon and then let it go. The air rushes out quickly, making the balloon shoot forward.

▶ Many machines, vehicles and tools need air in order to work. For example, air helps things to float in water. There is air inside the rubber tube round the sides of this dinghy. Without this air, the dinghy would sink.

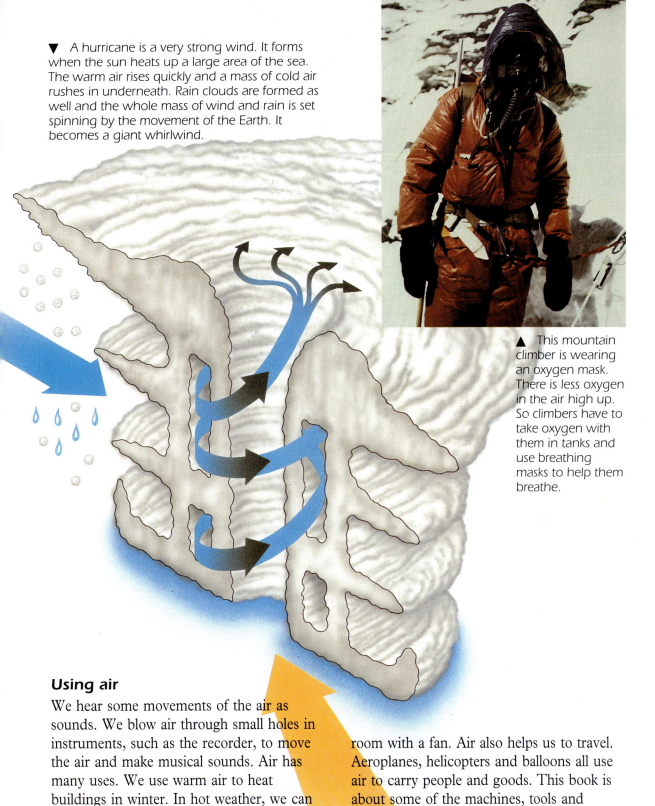

▼ A hurricane is a very strong wind. It forms when the sun heats up a large area of the sea. The warm air rises quickly and a mass of cold air rushes in underneath. Rain clouds are formed as well and the whole mass of wind and rain is set spinning by the movement of the Earth. It becomes a giant whirlwind.

▲ This mountain climber is wearing an oxygen mask. There is less oxygen in the air high up. So climbers have to take oxygen with them in tanks and use breathing masks to help them breathe.

Using air

We hear some movements of the air as sounds. We blow air through small holes in instruments, such as the recorder, to move the air and make musical sounds. Air has many uses. We use warm air to heat buildings in winter. In hot weather, we can cool a room by blowing cold air into the room with a fan. Air also helps us to travel. Aeroplanes, helicopters and balloons all use air to carry people and goods. This book is about some of the machines, tools and vehicles which use air, and how they work.

5

How your body uses air

We breathe in oxygen from the air and breathe out carbon dioxide into the air. This is called **respiration**. Many parts of our bodies take part in respiration. We call these parts the **respiratory system**.

The oxygen we breathe in joins up with substances in the food we eat. This changes the substances. The change makes heat and other forms of energy. We need energy in order to live and work.

Breathing air

You breathe in air through your nose and mouth. The air moves down your windpipe into tubes called **bronchi**. It then passes into the **lungs**. The lungs can hold about five litres of air. On average, each of us breathes in about 500 litres of air an hour.

The oxygen in the air passes from your lungs into your blood. Your **heart** pumps blood to your lungs and around your body. Your blood carries oxygen around your

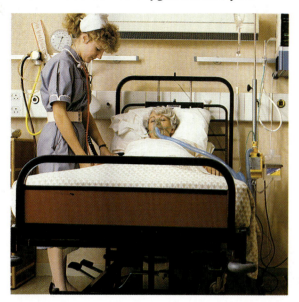

body. It also carries carbon dioxide back to your lungs. You get rid of this carbon dioxide when you breathe out.

Oxygen gives us the energy to do work, such as running, pushing and pulling. The harder we work, the more oxygen we need. This is why we pant when running. We breathe in and out more rapidly to suck extra oxygen into the lungs.

Extra air

People who live all the time in high mountains often have bigger lungs than normal. Their larger lungs help them to breathe in more air. They need to do this because there is less oxygen in the air higher up.

Mountain climbers sometimes carry extra oxygen to give them the energy they need at great heights. People who dive for a long time underwater have to carry all the air they need with them. They use special equipment called **scuba**. This usually consists of containers filled with air joined by tubes to a breathing mask. The diver breathes the air in through the mask, which covers the nose and mouth. People on land sometimes use equipment like this too. Fires use up the oxygen in the air. When there is a large fire it is hard for people to breathe. So fire fighters carry **breathing apparatus** which they use when they try to rescue people who have been caught inside a burning building.

◀ *Without oxygen we would die. This person's lungs are damaged. She cannot breathe in enough air to get the oxygen she needs. So this machine is giving her extra oxygen. When her lungs are healed, she will breathe normally again.*

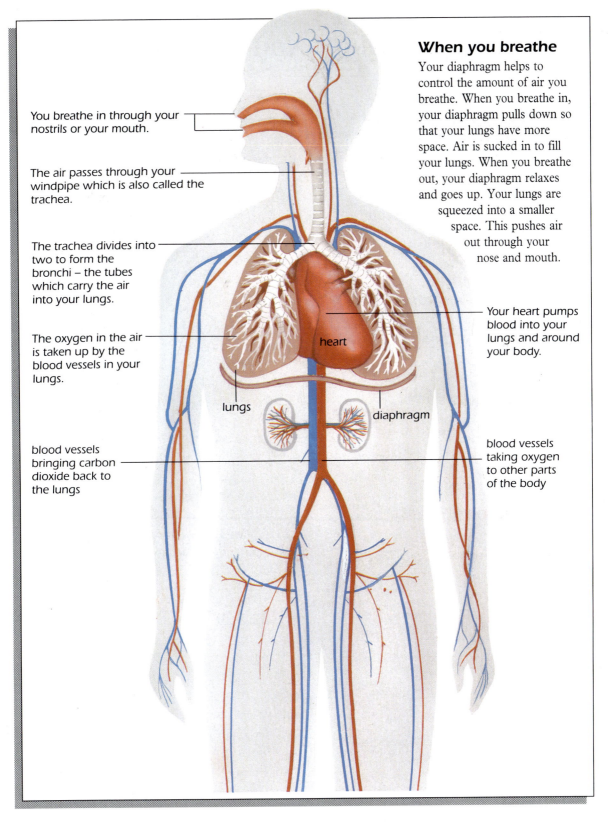

You breathe in through your nostrils or your mouth.

The air passes through your windpipe which is also called the trachea.

The trachea divides into two to form the bronchi – the tubes which carry the air into your lungs.

The oxygen in the air is taken up by the blood vessels in your lungs.

blood vessels bringing carbon dioxide back to the lungs

When you breathe

Your diaphragm helps to control the amount of air you breathe. When you breathe in, your diaphragm pulls down so that your lungs have more space. Air is sucked in to fill your lungs. When you breathe out, your diaphragm relaxes and goes up. Your lungs are squeezed into a smaller space. This pushes air out through your nose and mouth.

Your heart pumps blood into your lungs and around your body.

heart

lungs

diaphragm

blood vessels taking oxygen to other parts of the body

7

Making sounds

The air brings sounds to our ears. If you pluck a stretched rubber band you will hear a twang. This is because plucking the rubber band makes it move rapidly to and fro. This type of movement is called **vibration**. The movement of the rubber band makes the air around it vibrate. The vibrations spread through the air to your **eardrums**. They start to vibrate in a similar way. You hear these vibrations as a twang.

All the sounds we hear are made by vibrations. We hear slow vibrations as low sounds. We hear faster vibrations as higher sounds.

The effects of vibrations

We cannot see the vibrations in the air but we can sometimes see their effects. Some singers can make the air vibrate very fast to produce very high sounds. These air vibrations can be powerful enough to shatter a thin glass into tiny pieces.

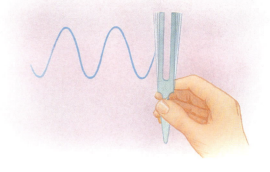

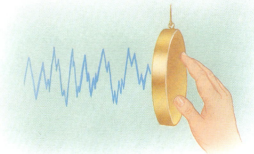

▲ We hear sounds when the air vibrates and makes our eardrums vibrate in a similar way. We can show the way the vibrations make the air move in a waveform. Some sounds have curved waveforms and some have jagged waveforms. The prongs on a tuning fork vibrate very regularly and have a curved waveform. Hitting a gong makes an irregular jagged waveform. We cannot see the waves but they cause changes in air pressure.

Blowing sounds

You can make the air vibrate just by blowing. If you cup your hands around your mouth you may be able to make a sound like an owl hooting. By tightening your lips so that you blow through a small round gap, you may be able to whistle. You can play a tune by humming through thin paper held in front of a comb. As you hum, the paper vibrates and makes the air vibrate.

◄► These people are both trying to make the sound of their voices travel further in the air.

Warning noises

Sirens are used to make loud warning sounds. Inside the siren is a disc or drum with holes in it. When you switch on the siren, it blows air on to the disc or drum. This spins rapidly so that bursts of air pass through the holes and make a sound. The faster the disc or drum spins, the higher the sound.

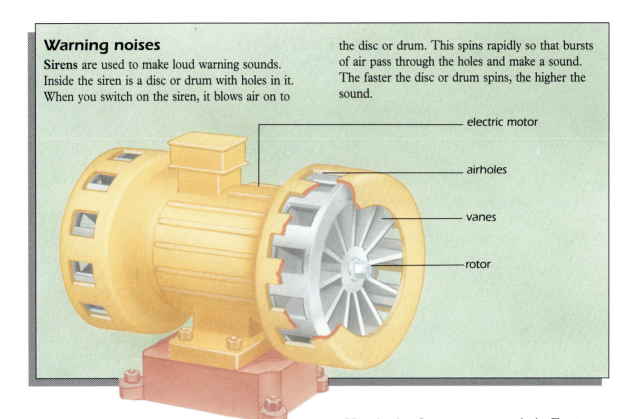

- electric motor
- airholes
- vanes
- rotor

All these sounds travel through the air to your ears. Sound can also travel through water and through solid objects, such as a wall. Most of the sounds we hear, however, come through the air.

Horns

When you shout at someone a long distance away, you probably cup your hands. This makes the sound go further. A hollow tube which is wider at one end makes your voice sound even louder. Tubes like this, called horns, were used to send messages or warnings to people far away. Over 2000 years ago, warriors blew horns to frighten their enemies. The horns they used were animals' horns.

Hundreds of years ago, people in Europe travelled in horse-drawn coaches. The guard on the coach blew through a horn to let people know that the coach was coming. Today, a car driver uses a horn to warn of danger. In a car horn, air is squeezed through a tiny hole to make a loud noise.

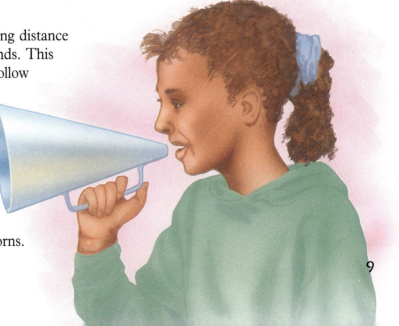

9

Making music

recorder

oboe

flute

clarinet

bassoon

▲ Brass instruments are made from metal. Woodwind instruments used to be made from wood. Now they are often made from metal or plastic as well.

A wind instrument makes music when the player blows air into a long tube through a special mouthpiece. This makes a column of air vibrate inside the tube. The high and low musical sounds made in this way are called notes. The longer the tube, the deeper the note. Instruments which work like this are either woodwind instruments, like the flute, piccolo and recorder, or brass instruments, like the trumpet and tuba.

Woodwind instruments

A woodwind player changes the length of the column of air in the instrument by covering the air holes. This is done with the ends of the fingers or by pressing keys. The shorter the column of air, the higher the note. When all the holes are covered, the instrument produces its lowest note.

The flute player and the piccolo player blow across the side of a hole at the top of the tube. The air from the player's lips strikes the sharp edge of the hole. This makes the air inside vibrate.

trumpet

trombone

tuba

saxophone

◄ The total length of tube in a brass instrument would be far too big to hold in the hand if it was stretched out straight. A trumpet has a tube almost 1.5 metres long. A straightened-out tuba would be six metres long! This is why the tubing is wound round and round like a snail.

The French horn

The French horn is also known as the double horn. This is because by opening the fourth valve (the thumb valve) a player can switch

between two sets of tubing. The longer tube is used to make a deep sound, the shorter tube is used to reach high notes.

Brass instruments

Brass instruments do not contain reeds. The trumpet player's lips vibrate like the reeds of the oboe. The vibrating lips and the shape of the mouthpiece make the air vibrate inside the tube of the instrument. The players of the trumpet, tuba or French horn press keys to make the column of air longer or shorter. The keys open and close valves. The valves direct the flow of air to the different lengths of tube inside the instrument. The trombone works in a different way. A trombonist changes the length of the tube by sliding it backwards and forwards.

A recorder player blows into a mouthpiece at the top of the tube. An oboe player blows between two reeds in the mouthpiece. These are pieces of thin tube made of cane. There is only one reed in the mouthpiece of a clarinet.

Moving air

Moving air can help to keep us cool in hot weather. On a very hot day people may try to find a breeze, or fan themselves with their hands or a newspaper. They try to make the air move.

You can test for yourself that moving air feels cool. Breathe onto your hands. Can you feel the warmth? Your breath feels warm because it is at a hotter **temperature** than your hands. Now blow hard through your fingers. Your breath feels cool, not warm, even though it is still at the same temperature. This is because the air is moving. The faster the air moves, the cooler it feels. Frosty days feel much colder when there is a wind than when it is calm.

Fanning the air

We use fans to make a cooling breeze when the weather is hot. In the past, fans were moved by hand. Today many fans contain an **electric motor**. This changes the energy of **electricity** into motion. The motor turns a rod called a **shaft**. There are fan **blades** fixed to the shaft. When the fan is switched on, the motor turns the shaft around. This makes the blades spin around. The blades are shaped so that they pull in air at the back and push it forward at the front to make a breeze.

Some fans hang down from the ceiling of a room. They have long blades which move quite slowly and push the air downwards. Other fans stand upright and blow air across a room. These fans have shorter blades and are small enough to be carried from room to room. Their fan blades spin rapidly, so they have metal bars in front to stop the blades hurting people's fingers.

◀ This family is keeping cool in the hot weather with the help of an electric fan. Air is sucked in at the back of the fan and blown out at the front. Because the air coming out of the fan is moving fast, it feels cooler.

fumes and smells sucked in here

fumes and smells blown out here

electric motor

fan

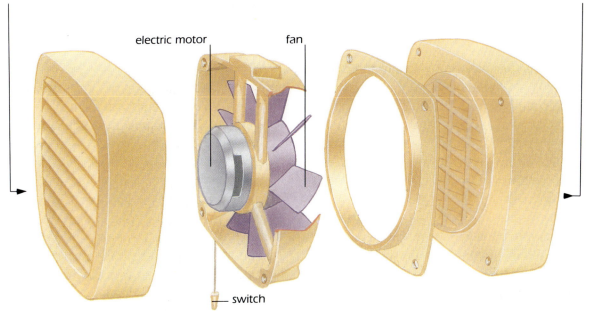

switch

▲ We can change the air in a room by using an extractor fan. This works in the opposite way to an ordinary electric fan. Instead of blowing air into a room, the extractor fan sucks out or extracts the old air. Fresh air takes the place of the old air. Extractor fans are often fitted in kitchens and bathrooms. They extract steam, fumes and smells, and let in fresh air.

Changing the air

We also use fans to bring fresh air into closed spaces. Inside a closed car, the air can become so stale that people cannot breathe properly. If the air inside a computer becomes too warm, the computer will not work. So cars and computers have fans. These replace the stale, warm air with fresh cool air. This changing of the air is called **ventilation**.

Road tunnels run under rivers and mountains. Cars give off **exhaust fumes** as they travel through the tunnels. If the fumes stayed in the tunnels, it would be hard for people to breathe. Giant fans change the air by blowing fresh air into the tunnels. The fresh air pushes out the fumes.

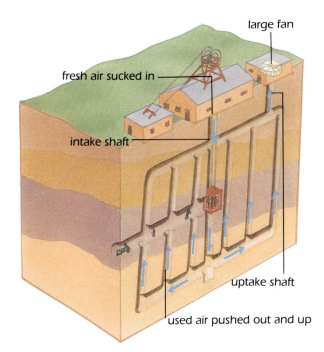

large fan

fresh air sucked in

intake shaft

uptake shaft

used air pushed out and up

▲ Fans provide fresh air for miners working underground. The mine has deep holes called shafts. The lift which takes miners from the surface to the mine travels down the intake shaft. A fan sucks fresh air down this shaft. The fresh air pushes the used air through another shaft to the surface.

Cooling with air

People feel most comfortable in a room which is not too hot and not too cold. It is most comfortable indoors if the temperature of the air is about 20°C. If the temperature falls below 15°C, people use heaters to make the air warmer. If the temperature of the air rises too much above 20°C, they try to make the air cooler.

The wetness of the air is also important. All air contains water vapour. If there is too much water vapour in the air, we feel sticky. If there is too little water vapour, our mouths and throats feel dry.

Many buildings have a system which keeps the air at the right temperature and wetness. This system is called **air conditioning**. It changes the condition of the air so that we feel comfortable.

▼ This diagram shows how air conditioning in a building works. The thermostat opens and closes valves to send the air through the heating unit or the cooling unit.

When it is hot

An air conditioning system has a **thermostat**. This device keeps the air at a fixed temperature. If the temperature gets warmer than this, the thermostat switches on the air conditioning system. A fan sucks the air in through a **filter**. Dirt, such as dust and smoke, is trapped in the filter, while the clean air passes through. It is then sucked through the cooling unit. This has tubes containing a cold liquid which takes in heat quickly. They take in heat from the air as it passes over them.

As the air cools, some of the drops of water vapour join together. The drops of water trickle down the tubes. So the cooling unit lowers the temperature of the air and makes it drier. The fan blows the cooler, drier air into the building through gaps in the walls called **vents**.

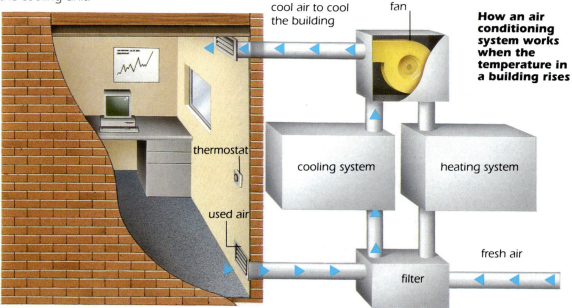

cool air to cool the building

fan

thermostat

used air

How an air conditioning system works when the temperature in a building rises

cooling system

heating system

filter

fresh air

◀ Air conditioning is needed in this library where many old and rare books are kept. If the air is too warm and dry, the pages of the books will become brittle. If the air is too cold and wet, the pages will become soft and tear easily.

When it is cold

If the temperature falls too low, the thermostat closes **valves** to shut off the cooling unit. Instead, the air is sucked through other valves into the heating unit. This has tubes containing liquid. The hot tubes give off heat into the air. Cold air is usually dry. So the air conditioner may also add water vapour to the air. The warmer, wetter air is again blown through vents into the building.

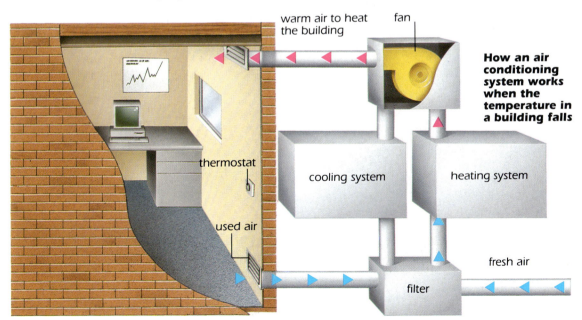

warm air to heat the building

fan

thermostat

used air

cooling system

heating system

fresh air

filter

How an air conditioning system works when the temperature in a building falls

Drying with air

Air conditioning alters the wetness of the air. We call the amount of water vapour in the air its **humidity**. On humid days your clothes feel damp and sticky as they take in water vapour from the air. Warm air can hold more water vapour than cold air. The tiny particles that make up the gases in the air move apart, or **expand**, as they get warmer. More water vapour can then be held between these particles. So warm days are often more humid than cold days.

▼ When you have a hot bath, you can see steam in the air. This is little drops of water called water vapour. When the steam meets the cold walls or windows of the bathroom, it condenses and turns back to water.

Water in the air

You can see the water vapour in the air condensing when you have a bath. The warm moist air in the bathroom cools when it touches cold surfaces, such as the windows. The water vapour condenses and

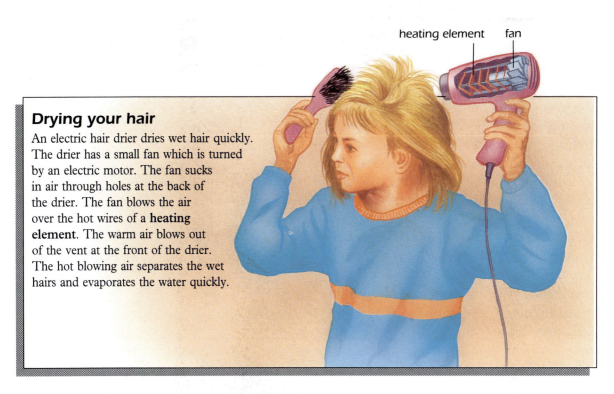

heating element   fan

Drying your hair

An electric hair drier dries wet hair quickly. The drier has a small fan which is turned by an electric motor. The fan sucks in air through holes at the back of the drier. The fan blows the air over the hot wires of a **heating element**. The warm air blows out of the vent at the front of the drier. The hot blowing air separates the wet hairs and evaporates the water quickly.

trickles of water run down the windows. When the air is warm and dry, the opposite happens. The air takes up more water vapour. Puddles of water dry up quickly, especially if the Sun is hot. Moving air also makes water evaporate. Wet clothes on a line dry quickly in a wind. When you rub yourself with a towel, evaporation helps to dry your body.

Drying machines

Some machines use warm moving air to dry up moisture. A tumble drier moves the clothes around gently so that the warm air blows on every part of the clothes. The water evaporates into the air leaving the clothes dry. The wet air goes through a vent to the outside of the building.

The drying machine in a washroom blows hot air on your hands. The machine dries wet hands in only a few seconds. Unlike a towel, it never gets dirty.

Water heated by the Sun

There is water vapour in the air because of the Sun. It warms the water in the seas, lakes and rivers. The Sun's heat makes some of the water turn to vapour. The water vapour rises up into the air.

We cannot see the water vapour in the air but we can see it when it turns back into a liquid. On clear summer nights, the air cools and can no longer hold as much water vapour. Some of the water vapour turns to water, or **condenses**. It forms tiny drops of **dew** on grass and other plants.

Sucking up dirt

When you drink through a straw, you make liquid flow upwards. Why does this happen? Before you suck, the water in the straw is at the same level as the water in the glass. This is because there is air in the straw and the glass. The air presses down on the water, keeping it at the same level. When you suck, you take out the air in the straw. So there is nothing pressing on the water in the straw. The air still presses on the water in the glass, pushing the water up the straw. This push is called **suction**.

If water did not take the place of the air in the straw, there would be an empty gap called a **vacuum**. Suction makes anything nearby move to fill a vacuum.

A vacuum cleaner

Many machines use suction to do work. A vacuum cleaner uses suction to clean up dust and dirt. Some vacuum cleaners stand upright. An upright vacuum cleaner has a brush wrapped around a tube called a roller. A rubber belt stretches around the roller and around the shaft of an electric motor. This moves the belt which turns the roller and brush around. As the brush turns it stirs up the dirt and dust on the floor. At the same time the shaft spins the blades of a fan behind the brush. The fan is at the front of a tube. The fan sucks air, loose dust and dirt into the tube, just as you suck water up a straw. The air and dirt passes through the tube into a paper bag, fixed to the upright handle. The dirt is trapped in the paper bag. The air passes through the paper and vents in the cleaner back into the room.

▼ When you drink milk through a straw, you suck the air out of the top of the straw. Air presses on the milk in the glass and pushes it up the straw.

The other main type of vacuum cleaner is shaped like a tube or **cylinder**. The cylinder vacuum cleaner lies on the floor. At one end, there is a long tube which can bend to go round corners. You can fit different brushes to the end of this tube. Usually, the brushes do not turn around. A powerful fan sucks up the dust and dirt through the tube into a paper bag. The air leaves through a hole in the other end of the cylinder.

Small cleaners

The large upright and cylinder vacuum cleaners must be plugged into a **power point** before they will work. When the cleaner is switched on, electricity flows from the power point to the electric motor. Smaller vacuum cleaners can use the electricity from a **battery**. These cleaners are much easier to carry around.

tube

paper bag

brush roller

electric motor

fan

rubber belt

▲ The vacuum cleaner works by suction. The fan draws air and dirt into the cleaner. The dirt collects in the bag and the air escapes through the vents.

19

Pumping up

Like the vacuum cleaner, a **pump** works by suction. A simple pump is a cylinder containing a thick rod or disc. It fits tightly against the walls of the cylinder so that not even air can get around it. The rod or disc is called a **piston**. The piston can be moved up and down the cylinder.

The bottom of the cylinder is placed in a substance, such as water. Then the piston is pulled up. This pulls the air up to the top of the cylinder leaving a vacuum. Suction makes the water rise up into the bottom of the cylinder to fill the vacuum.

▲ This bicycle pump uses a piston in a cylinder to pump air into the tyre.

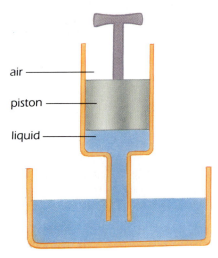

◄ This is how a simple pump works. As the piston is pulled up, the liquid moves up into the cylinder. The piston has to fit very tightly inside the cylinder so that air does not get down the sides.

A simple pump

When you have an injection, the nurse uses a simple pump. It is called a **syringe**. The top of the piston is at one end of the syringe. At the other end there is a needle with a hole up the centre. The nurse places the needle in a liquid and pulls up the piston. This sucks the liquid through the needle and into the tube. Then the nurse puts the needle in your arm and pushes the piston down. The liquid goes out through the needle and into your arm.

Pumping air

When you put air into a bicycle tyre, you use a simple pump. When you pull up the top of the bicycle pump, you pull the piston up the cylinder. As the piston rises, suction opens a valve at the bottom of the cylinder. Air is sucked in through the valve. When you push the pump handle in again, the piston moves back down the cylinder. It pushes against the air which forces open another valve. This valve is in the middle of the piston. The air flows through this valve into the cylinder above the piston. When the piston reaches the bottom of the cylinder, the middle valve in the piston closes. Then the piston rises up the cylinder again. The **compressed** air beneath the piston is pushed out through another valve at the bottom of the pump into the tyre.

Other pumps

You see many types of pump every day. Most **central heating** systems have pumps to send hot water through pipes and radiators. Pumps push water into washing machines and dishwashers. Fire brigades use pumps when they fight a fire. You also carry a pump around with you every day. Your heart pumps blood around your body.

Centrifugal pumps

Some pumps spin blades around at high speed inside a round box. These are called **centrifugal** pumps. They suck air up or move liquids and loose solids such as grain.

▼ The petrol pumps at a filling station suck up petrol from a huge tank underground. The pump spins around as it sucks up the petrol. This is why it is called a rotary pump. The pump has a meter to measure the amount of petrol each customer takes. The petrol flows through the meter. The customer controls the amount of petrol flowing into the car by pressing the lever on the handle.

meter

hose

rotary pump

tank underground

21

Drilling down

When you work a bicycle pump, you give the air inside enough power to push into the tyre. You do this by squeezing or compressing the air into a small space. The more air is compressed, the more power it has. Air can even be made powerful enough to enable a machine to drill through hard rock. We use **pneumatic drills** in this way to break up the road surface. Compressed air moves a piston up and down a cylinder inside the drill. At the bottom of the drill is a tough, sharp blade which rests on the road surface. When the air presses the piston down, it strikes the blade. The piston hammers the blade into the road and breaks up the surface. Pneumatic drills are also known as jack-hammers.

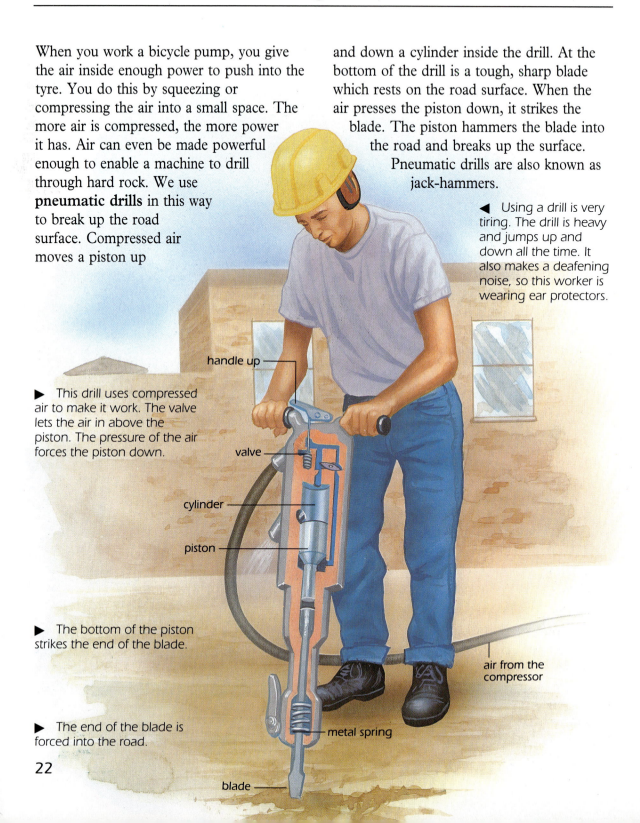

◀ Using a drill is very tiring. The drill is heavy and jumps up and down all the time. It also makes a deafening noise, so this worker is wearing ear protectors.

▶ This drill uses compressed air to make it work. The valve lets the air in above the piston. The pressure of the air forces the piston down.

handle up

valve

cylinder

piston

▶ The bottom of the piston strikes the end of the blade.

air from the compressor

metal spring

▶ The end of the blade is forced into the road.

22

blade

How a pneumatic drill works

The air used by a road drill is compressed in a machine called a compressor. The road worker presses a handle at the top of the drill to let in the compressed air. It shoots into the drill with great force. At the top of the cylinder inside the drill is a valve. It can rock backwards or forwards, letting air in below or above the piston. First, the valve rocks back so that the air goes below the piston and pushes the piston up the cylinder. As the piston is pushed up the cylinder the air trapped above it pushes against the valve. The valve rocks forwards and allows the air in above the piston. The air drives the piston downwards with such force that it hammers down on top of a strong blade. The piston drives the blade into the ground. The valve

▲ *This miner is drilling for coal using a pneumatic drill.*

then rocks back and the air pushes the piston back up the cylinder. As it does so, a strong metal spring pushes the blade back up. It is now ready to be hammered again. These blows are repeated over and over again, and the drill breaks up the road surface.

The pneumatic drill

The pneumatic drill was invented over 125 years ago, by an engineer working on the Mont Cenis railway tunnel. The tunnel runs under the Alps between France and Italy. Using drills, the workers dug the tunnel three times as fast as they would have done using hand tools.

Cushions of air

▲ This airbed can support a person and float on water. It is filled with compressed air. The armbands are filled with air too. They help this little boy to float while he learns to swim. When the beach ball has plenty of air inside, it bounces very well. If the air escapes, it becomes softer and will not bounce.

We can ride on air too. Rubber tubes filled with air put a cushion between the ground and different kinds of vehicles. The tyres of bicycles, buses, trucks, cars and aircraft are all filled with air.

A rubber tyre

A hundred years ago, vehicles had solid wheels made of wood, metal or rubber. A journey in the vehicles was hard and bumpy. Then an Irish inventor called John Boyd Dunlop tried to make his son's tricycle more comfortable. Round each wheel, he fitted an outer canvas cover. Inside the cover was a rubber **inner tube** filled with air. Riding on tyres filled with air is softer, smoother and quieter than riding on solid tyres.

▶ The rubber tubes inside bicycle tyres need to be strong to keep in the compressed air. If you get a puncture, you can mend it by sticking a thick rubber patch over the hole.

A few years later in France, the Michelin brothers used air-filled or **pneumatic tyres** on motor cars. These early pneumatic tyres were not very strong. At that time there were many more horses than cars on the roads. Old horseshoes and nails lying in the road often punctured a tyre's inner tube, letting the air out. The outer cover was made stronger but the inner tube was still sometimes punctured. About 40 years ago, a new type of tyre was invented, which is much more difficult to burst. It has no inner tube and so is called a **tubeless tyre**.

Inside a tyre

The tubeless tyre has a thick coating of rubber covering several layers of tough cord called **plies**. These give the tyre strength. The plies are built up in layers. The edges of the tyre contain steel wire to make them firm. The edges are fixed so tightly to the rim of a wheel that air cannot escape. The tyre is filled with compressed air. A pump forces the air into the tyre through a valve.

This valve lets air into the tyre but stops the air already inside from escaping.

A car tyre protects the passengers from bumps in the road. The thick rubber coat also helps to stop the car sliding over slippery roads. The rubber has a deep pattern cut in it called the **tread**. This helps the tyre to grip the surface of the road and helps to move water from under the tyre.

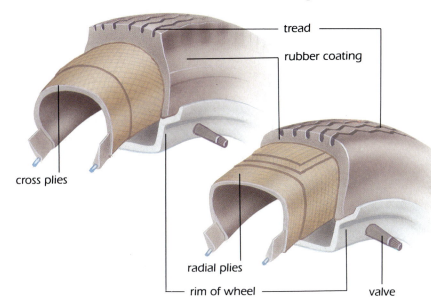

tread

rubber coating

cross plies

radial plies

rim of wheel

valve

◀ The tyres on cars are made of several layers of rubber and metal. They usually do not have tubes. The layers, or plies, are laid in different directions. Some are called cross-ply tyres and others are called radial-ply tyres. Can you see how they are different?

Riding on air

Machines without wheels can also ride on air. Some lawnmowers have no wheels, and ride on a cushion of air. A fan draws in air and pushes it down through a hole in the bottom of the mower. This compresses the air below the mower. The compressed air lifts the mower so that it **hovers** above the ground. A mower like this is a type of **hovercraft**. Since it does not touch the ground, a hovercraft is easy to move. It can carry passengers and cars. Hovercraft can travel over land but are usually used for journeys across water.

▶ A hovercraft rides on a cushion of air. The air is taken in through the air intakes and pushed down under the craft where the flexible skirt keeps it in place. When the hovercraft lands, it sinks down as the air is released. A typical hovercraft can carry 250 people and 30 vehicles and travels at up to 90 kilometres per hour.

Hovercraft

A hovercraft can travel much faster than most boats. This is because it is much easier to move through the air than through water.

Powerful fans suck in air through holes, called air intakes, on the upper deck. The air is compressed into a small space below the passenger deck. The compressed air escapes through holes around the edge of the hovercraft. There is a rubber or plastic skirt wrapped around the edge. The skirt keeps the air in so that it forms a cushion under the hovercraft. The air lifts the hovercraft about half a metre above the surface of the sea.

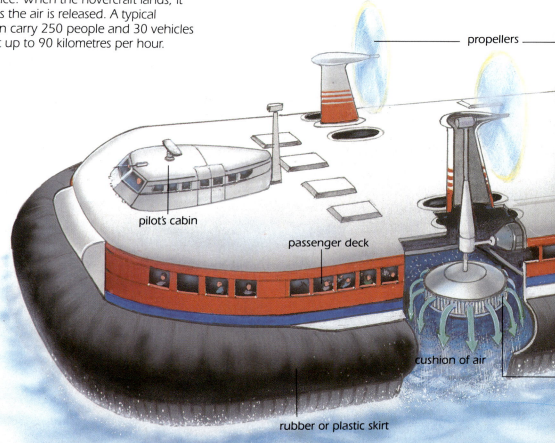

propellers

pilot's cabin

passenger deck

cushion of air

rubber or plastic skirt

◄ Many vehicles can travel as fast over good land as the hovercraft. However, it can travel over difficult or dangerous land such as a swamp, which most other vehicles cannot cross.

Each fan is driven by a large engine. Each engine also turns large blades called **propellers.** As the propellers spin, they move the air which pushes the hovercraft forward.

The pilot of the hovercraft sits in a cabin at the front of the top deck. The pilot steers the hovercraft by moving huge **rudders**. These change the direction of the air at the back of the craft.

Skimming over the sea

A hovercraft skimming across the sea is exciting to look at. It is not so exciting for the passengers inside. The seats are close together like those in an aeroplane. There is little to see but water and spray on the windows. A journey in a ship is much more interesting.

People travel by hovercraft because it is much quicker than a ship. However, travelling by hovercraft is also more expensive. Hovercraft engines use a lot of **fuel**. So hovercraft are mainly used on short journeys, such as across the English Channel. The fuel needed for long journeys would take up so much space there would be little room for passengers or cars.

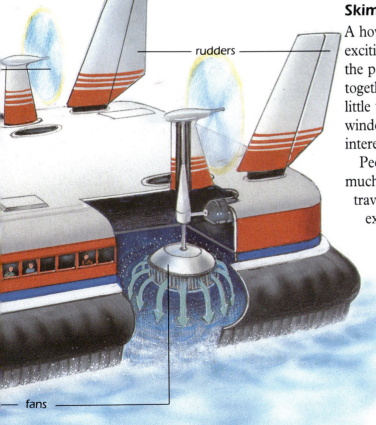

rudders

fans

Sailing with the wind

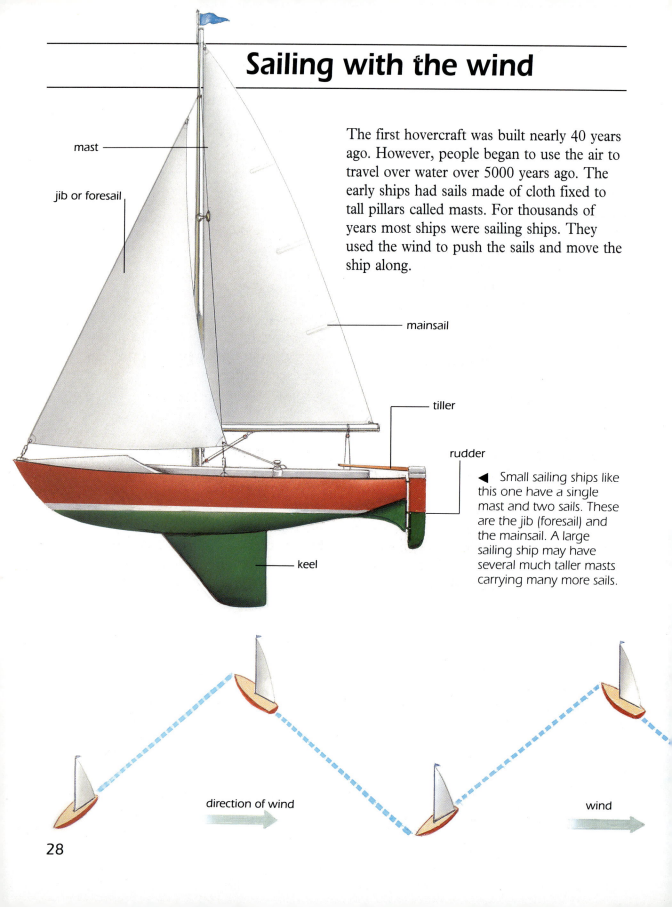

mast

jib or foresail

The first hovercraft was built nearly 40 years ago. However, people began to use the air to travel over water over 5000 years ago. The early ships had sails made of cloth fixed to tall pillars called masts. For thousands of years most ships were sailing ships. They used the wind to push the sails and move the ship along.

mainsail

tiller

rudder

◄ Small sailing ships like this one have a single mast and two sails. These are the jib (foresail) and the mainsail. A large sailing ship may have several much taller masts carrying many more sails.

keel

direction of wind

wind

Using the wind

The pressure of the wind on the sails pushes the ship through the water. The sailors steer the ship in the right direction. They do this partly by moving the sails round. You can test how this works for yourself. Bend a small piece of card in two and stand it like the sides of a tent on a smooth table. Place the card so that one side is facing you. If you blow hard, the card will move straight away from you. Now move the card so that the side is slanted away from you. When you blow, the card moves at a slant instead of straight ahead.

The sailors can make the ship move forward even when the wind is blowing from the front of the ship. To do this, the sailors turn the sails to one side and then to the other. This makes the ship move along a zig-zag course. It is called **tacking**. However, the sailors cannot sail the ship when there is no wind. On calm days there is no moving air to push the sails and the ship cannot move. This is why many modern sailing ships also have an engine to use on calm days.

▲ This is a modern Japanese tanker. It saves fuel by using sails as well as engines. The sails are moved by a computer to make the best use of the wind.

A sailing ship

A crew of sailors controls the speed and direction of the ship. They do this by changing the position of the sails. The sailors also use the rudder to help steer the ship through the water. They move the rudder by pushing a rod called a tiller.

The underside of the sailing ship, called the keel, is heavy. Objects with a heavy bottom are difficult to tip over. The keel keeps the ship upright even when the wind is pressing very strongly on the sails.

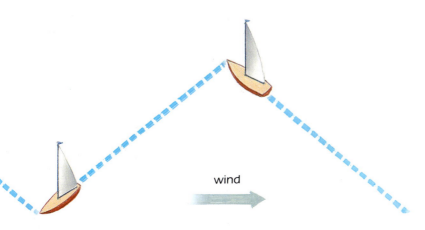

wind

◀ Sailing boats can move in any direction provided there is a wind. To sail into the wind, the sailors turn the sails so that they are at an angle to the wind. Turning the sails to one side and then to the other makes the ship move forward on a zig-zag course.

Working with the wind

Over 2000 years ago, people started using the wind to work machines, as well as to move ships. People on the Greek island of Crete were among the first to build windmills. They had sails like those on a sailing boat. The wind turned the sails.

The windmill's sails were fastened to long crossed poles. These were fixed in the middle to a wooden shaft at **right angles** to the poles. When the wind blew, the sails turned the shaft around.

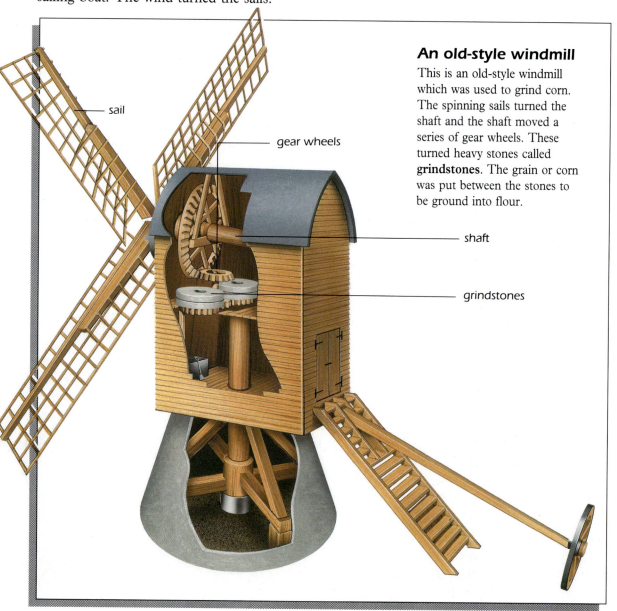

sail

gear wheels

An old-style windmill
This is an old-style windmill which was used to grind corn. The spinning sails turned the shaft and the shaft moved a series of gear wheels. These turned heavy stones called **grindstones**. The grain or corn was put between the stones to be ground into flour.

shaft

grindstones

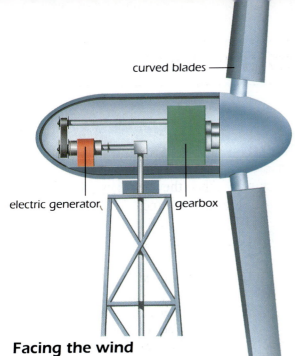

curved blades

electric generator   gearbox

◀ Modern windmills are much more powerful than the windmills of the past. Wind generators like this one have huge blades instead of sails. The blades can be up to 100 metres across. The sails drive a generator for making electricity. A computer makes sure that the blades always face the wind.

Facing the wind

A windmill's sails only spin when they are facing the wind. When the wind changes direction, the sails stop moving. So people built windmills with sails which could be turned around. Every time the wind changed direction, the sails had to be turned to face the wind again. About 200 years ago, people found a way to make the sails turn to face the wind **automatically**. A small ring with blades, called a fantail, was fixed at right angles to the sails. When the wind changed direction, it blew against the fantail, moving it around. The fantail turned gear wheels. These moved the sails around until they faced the wind again.

Modern windmills

All windmills have one big disadvantage. They cannot work when there is no wind. Electricity and fuels, such as petrol, can give power all the time. Today, most of our machines are powered by electricity or fuels. However, these are expensive, while the power of the wind is free. So we still use some windmills. Windmills are used to pump water out of the ground in many parts of the world. Other windmills, called wind **generators**, use wind power to make electricity. Instead of sails, the wind generators have long blades like propellers. The blades are curved so that they catch the air. When the wind blows, the blades spin around much faster than the sails of an early windmill. As they spin, the blades turn a generator which makes electricity. Some wind generators can even make electricity when there is no wind. They use the extra electricity they make on very windy days to pump water into a special lake above the windmill. On calm days, the water is released from the lake and falls down onto a turbine. The falling water turns the turbine to make electricity.

▼ In windy places, people sometimes build many wind generators close to each other. They are called wind farms, and can make a lot of electricity. This wind farm is at Altamont Pass, California.

Balloons and airships

◄ Modern hot-air balloons are often used for advertising. They sometimes appear at fetes and fairs and you may be able to buy a ride in the basket. It takes about 15 minutes to heat the air in a hot air balloon and make it rise. Modern burners use propane gas as a fuel. They heat the air in the balloon to about 100°C. When the crew want the balloon to go higher they light the burner again.

An air bubble in a bowl of water rises quickly to the surface. This is because the air bubble is much lighter than the water. A gas called **hydrogen** is lighter than air. So a balloon filled with hydrogen rises up through the air.

The first balloons

On 21 November 1783, in Paris, a balloon carried people up into the sky. It was filled with air heated by burning paper. Hot air is lighter than cold air and so the balloon rose into the sky.

Ten days later another balloon was ready to fly above Paris. This balloon was filled with hydrogen. Heavy bags of sand in the basket kept the balloon on the ground. To make the balloon rise, the crew threw out some of the sand. When the crew wanted to come down again, they let hydrogen out of the balloon. However, the crew could not make the balloon move to the right or left. It was just blown along by the wind.

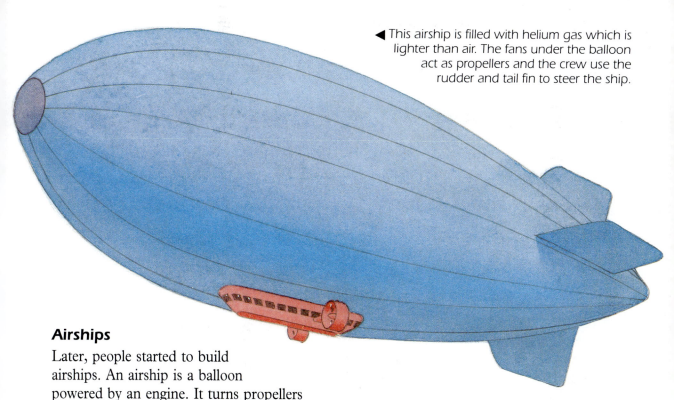

◀ This airship is filled with helium gas which is lighter than air. The fans under the balloon act as propellers and the crew use the rudder and tail fin to steer the ship.

Airships

Later, people started to build airships. An airship is a balloon powered by an engine. It turns propellers which push the airship through the air. The crew can steer the airship using a rudder.

About 90 years ago in Germany, Count Ferdinand von Zeppelin built huge airships, called Zeppelins, shaped like giant cigars. The crew and passengers sat in a cabin filled with air.

The airship also contained vast bags filled with hydrogen. However, hydrogen is very dangerous. A spark or flame can set hydrogen on fire. It burns so quickly that it explodes. In 1937 an airship exploded, killing many people. Modern airships use a safe light gas called helium instead of hydrogen.

Using hot air

Air is much safer than hydrogen. Today, most of the balloons used to carry people are filled with hot air. A hot air balloon has a large opening at the bottom. **Gas burners** heat the air, which rises through the opening and fills the balloon.

Turning up the gas burners makes the air hotter, and the balloon rises up into the sky. This happens because air takes up more space as it gets hotter. So some air is pushed out of the balloon. This makes the balloon lighter and it rises up. Turning the gas burner down makes the air cooler. The air shrinks and more air is sucked in, so the balloon becomes heavier. This makes the balloon move down towards the ground.

▶ This balloon is carrying instruments to measure the wind and the temperature of the air. It is being used by scientists who study the weather. The balloon is filled with hydrogen.

33

Gliding and floating

Birds can glide and float through the air. Yet birds are heavier than air. For hundreds of years people tried to discover their secret. Some people built wings and tried to fly through the air. Instead, they fell to the Earth like a stone.

This was because the Earth pulls everything down towards it. This pull is called **gravity**. If you throw a ball up into the air, the ball falls back to the ground, pulled by gravity.

A bird, however, can make the air push up under its wings. When this push is stronger than the pull of gravity the bird can rise up in the air.

► Modern parachutes have holes which let out air. The escaping air pushes the parachute along. The parachutist can control the direction in which the parachute moves. This is why a parachutist can land on a target on the ground.

panel

main parachute

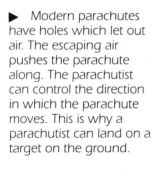
quick release hooks

harness

Down to Earth

In 1783 Louis-Sebastien Lenormand floated through the air like a giant bird. He hung from a huge piece of material called a **parachute**. It was shaped like the top half of a balloon. The air trapped inside the parachute pushed upwards. This push was not as strong as the pull of gravity so he fell down. However, he fell very slowly and landed safely.

Today, people use parachutes to help them fall safely from great heights. Parachutists often jump from planes high up in the sky. The parachute is folded up in a pack on the parachutist's back. The parachutist drops down through the air like a stone for hundreds of metres. Then the parachutist pulls a cord called the ripcord. This pulls open a small parachute at the top of the pack. The air pushes open this **pilot parachute** quickly. It tugs the main parachute up out of the pack. Seconds later the main parachute is full of air, slowing the fall of the parachutist.

Like a bird

Many people use **hang gliders** to fly for pleasure. Most modern hang gliders have two wings, side by side. The pilot hangs from the wings held by straps called a harness. The pilot can move the wings to steer the hang glider.

The pilot of a hang glider uses the fact that warm air is lighter than cold air. There is warm air above the ground on the slopes of steep hills and mountains. The warm air rises up in streams called **thermals**. These are like winds which blow upwards instead of along the ground. You can sometimes see thermals among huge clouds in summer. Look for parts of the cloud climbing rapidly in the sky.

The thermals push the wings of the glider up into the sky. Hang gliders can climb to great heights in this way. A skilful pilot can use the thermals to fly through the air for several hours at a time.

▼ A hang glider pilot can steer the glider by simple body movements. To move to left or right, the hang glider pilot shifts his body weight to left or right. By pulling his body forward, the pilot can increase the speed of the glider. To slow down, the pilot pushes backwards. This changes the angle of the glider.

Flying an aircraft

Like birds and gliders, aircraft need wings to raise them up into the air. The air moving round the wings pushes them up. This push is called **lift**. If there was no air, birds, gliders and aircraft could not fly.

The secret of an aircraft's flight is in the shape of its wings. Each wing has a flat underside. The upperside is curved so that the wing is rounded at the front edge and tapers to a point at the back. This shape is called an **aerofoil**.

Take-off

To take off, the aircraft travels at high speed along a runway. As it does so, the air rushes by. When it reaches the wings of the aircraft, the air is split in two. Some of the air passes above the wing and some passes below. The wing's curved top is larger than the flat underside. So the air above the wing has to travel further and faster than the air below. This makes the pressure of the air below the wing greater than the pressure of the air above. The difference in pressure gives the aircraft lift.

How an aircraft flies

The lift of the wings has to be strong enough to stop gravity pulling the aircraft down. The pull of gravity on an object is called its weight. When an aircraft's lift is greater than its weight, it will take off.

An aircraft is pushed along a runway and through the air by engines. This push is called **thrust**. In a propeller-driven aircraft engines also turn the propellers around. These suck in air at the front of the aircraft. The air moving backwards at high speed pushes the aircraft forwards.

Thrust helps the aircraft overcome another force, called **drag**. This is the push from the air which you can feel when you cycle fast. Drag holds you back, especially if there is a wind against you as well.

▼ Heavy aircraft need a strong lift to raise them off the ground. The lift of the wings increases with the speed of the aircraft. So heavy aircraft have powerful engines and take off from long runways. This gives them the speed they need to get enough lift.

▶ The shape of an aircraft's wings provides the lift it needs to take off.

The pilot can move parts of the tail and wings to help steer the aircraft in the sky. On either side of the tail are **elevators**. Moving these up or down makes the aircraft dive or climb. In the middle of the tail is the rudder. This turns the aircraft to the left or right, which is called yawing. On the back edge of the wings are **ailerons**. Moving these makes the aircraft **bank** as it turns. Banking stops the passengers being thrown from their seats.

▼ The flow of air above the curved part of the wing has further to travel than the air below the straight edge.

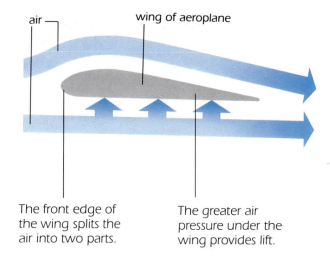

air

wing of aeroplane

The front edge of the wing splits the air into two parts.

The greater air pressure under the wing provides lift.

▼ The arrows in the picture show the three ways in which aircraft can move. Banking is achieved by yawing the aircraft and rolling it inwards at the same time.

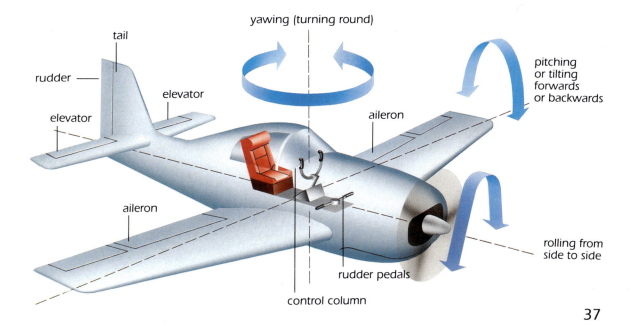

yawing (turning round)

tail

rudder

elevator

elevator

aileron

pitching or tilting forwards or backwards

aileron

aileron

rolling from side to side

rudder pedals

control column

Flying a helicopter

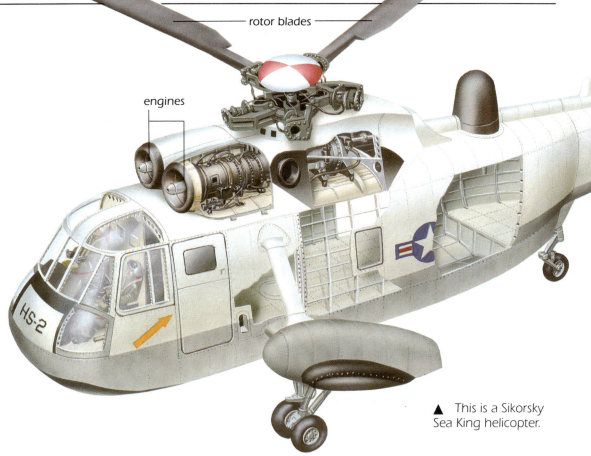

rotor blades

engines

HS-2

▲ This is a Sikorsky Sea King helicopter.

A helicopter does not have to move quickly along a runway in order to take off. It can rise straight up into the air and come straight down to land. This is a big advantage. Planes must take off from runways, such as those at airports. To journey by plane, passengers have to travel to and from an airport. A helicopter can take passengers from the roof of one building to the roof of another. So short journeys are much quicker and easier by helicopter than by plane.

▶ Helicopters are often used when there is a disaster or a serious accident. A helicopter can land almost anywhere. If it cannot land, the helicopter hovers. People can be pulled up to the helicopter using ropes and ladders.

tail rotor

◀ As the rotor blades spin, they push the body of the helicopter around in the opposite direction. The tail rotor spins to stop the helicopter being turned around.

Rotating wings

Like a plane, a helicopter has aerofoils to give it lift. Aerofoils must be moving quickly through the air in order to give lift. The aerofoils of a helicopter move around (rotate) quickly while the body of the helicopter stays still. This is why a helicopter can rise straight up off the ground.

The aerofoils of a helicopter are called **rotors**. They are carried on top of the helicopter, like the blades of a giant fan.

▼ A helicopter can fly backwards, forwards and sideways.

Powerful engines turn the rotor blades. As they rotate, they make the pressure of the air under the rotor blades greater than the pressure of the air above. This gives the lift which pushes the helicopter straight up into the air. The faster the blades turn, the greater the lift.

Controlling a helicopter

The pilot can make the helicopter travel in every direction and at different speeds. The pilot controls the helicopter by changing the speed and position of the rotors. When the rotors are level, the helicopter will move up or down. Increasing the speed of the rotors increases the lift and makes the helicopter climb. Decreasing the speed of the rotors makes the helicopter descend. When the lift from the rotors pushing upwards is the same as the weight of the helicopter pulling down, it hovers.

The rotors can give the helicopter thrust as well as lift. By tilting the rotors, the pilot can steer the helicopter in any direction. When the rotors are tilted to the front, the helicopter moves forward. Increasing the speed of the rotors makes the helicopter move forward faster. The helicopter can also fly backwards, and from side to side.

flying upwards: rotor blades level and spinning fast; lift more powerful than weight

flying backwards: rotor blades tilted backwards

flying forwards: rotor blades tilted forwards

hovering: rotor blades level; lift the same as weight

flying sideways: rotor blades tilted sideways

Jets of flame

Jet aircraft get their thrust from jet engines. These work in a different way from the engines which turn propellers. A jet engine produces gases which shoot out of the back of the engine. For every action there is an equal and opposite action. So, the gases shooting out backwards make the engine and the aircraft move forwards.

A jet engine

There are several types of jet engine. All push out gases to give aircraft the thrust they need to fly. A jet engine contains a machine with rapidly turning blades called a **turbine**. This sucks in air at the front of the engine. The air is squeezed in a compressor. Then liquid fuel is sprayed into the compressed air through a **nozzle**. The fuel is mixed with air

because fuel needs the oxygen in air in order to burn. The mixture of compressed air and liquid fuel is set on fire. It burns so strongly that the liquid fuel changes into hot gases. As the gases heat up, they expand to fill the engine. There is not enough room in the engine for the compressed air and hot expanding gases. So they shoot backwards out of the engine as powerful jets of flame. This gives the aircraft a powerful thrust forwards. Some aircraft have an even more powerful jet engine called a **turbofan**.

▼ A jet engine burns a mixture of compressed air and fuel. The fuel changes into hot gases which rush out at the back of the engine. This provides the thrust to lift the aircraft. Turbofan jet engines, like this one, have a huge fan at the front. This makes the air and gases move faster, giving the engine an even greater thrust.

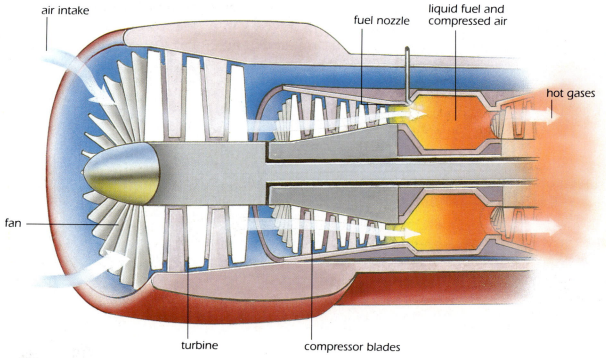

air intake

fuel nozzle

liquid fuel and compressed air

hot gases

fan

turbine

compressor blades

◄ The hot gases that shoot out of the back of the jet engine provide the thrust for a jet plane's take-off.

Flying a jet aircraft

A jet engine can give a much more powerful thrust than propellers. So a jet aircraft can fly much faster than propeller-driven aircraft. Most of the aircraft used today have jet engines. The fastest aircraft are the **supersonic** jet planes. Like all aircraft, a jet plane needs aerofoils to give it lift. The wings of supersonic jet planes are often smaller than those of other aircraft. Fast aircraft also often have swept back wings.

How a jet engine works

You can see how a jet engine works if you blow up a balloon and release it. The balloon moves forward rapidly. This is because when you blow air into the balloon, you compress the air. While you hold the balloon closed the air cannot escape. When you let go of the balloon, the compressed air rushes out. The air moves backwards at great speed. This makes the balloon move in the opposite direction.

The world's fastest passenger aircraft, Concorde, has wings shaped like a triangle. Swept back wings lessen drag and so help the aircraft move quickly through the air.

► Supersonic aircraft take off and land at such high speeds that small wings can lift them up.

Rockets into space

A rocket engine works in a similar way to a jet engine. Hot expanding gases give the rocket a powerful thrust. The gases from the rocket engine shoot downwards. This makes the rocket shoot up away from the Earth and into space.

As the rocket moves upwards, gravity is pulling the rocket back down to Earth. The rocket must travel very fast in order to escape the pull of the Earth. The rocket must reach a speed of over 30 000 kilometres an hour, much faster than a supersonic aircraft. So rocket engines have to be very powerful.

How a rocket moves

Most rockets have no wings or propellers, which use the air in the atmosphere to help aircraft fly. In space there is no atmosphere. So rocket engines must provide all the power for a rocket.

A rocket has many engines, each of which burns fuel. Fuel needs oxygen in order to burn. Since there is no air in space, the rocket must carry oxygen with it. To make the oxygen easier to carry, it is compressed into a small space. This makes the oxygen change from a gas into a liquid.

Many rockets have three parts or stages. There are engines in each stage. The first stage is the most powerful. It lifts the heavy rocket off the ground and sends it streaking

▶ This is a Saturn V rocket which was used by the Apollo missions to the Moon. You can see the three stages of the rocket, each with its own engines. The spacecraft, where the astronauts lived, is right at the top of the rocket. Just below is the lunar module in which two astronauts descended to the surface of the moon.

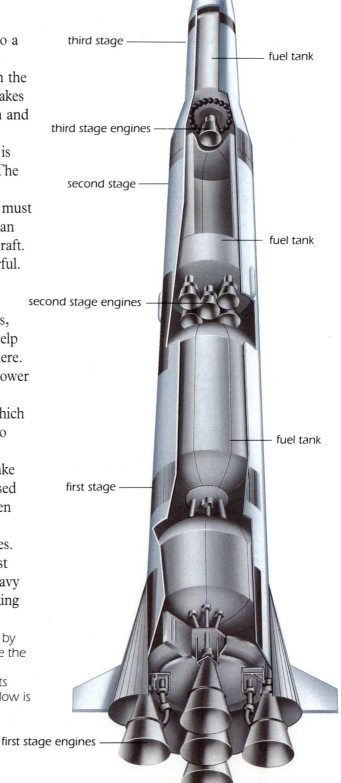

third stage

fuel tank

third stage engines

second stage

fuel tank

second stage engines

fuel tank

first stage

first stage engines

into the sky. After about two minutes, all the fuel in the first stage has been burnt. The first stage is pushed away, and this makes the rocket lighter.

By this time the rocket is travelling at over 6000 kilometres an hour. The second stage rocket engines take over. About three minutes later, the second stage engines fall away. Now the rocket is much lighter and is travelling at 17 000 kilometres an hour. The third stage engines shoot the rocket into space at a speed of over 35 000 kilometres an hour.

The Space Shuttle

Rockets are very wasteful. Each stage costs a lot of money to build. The first two stages drop off in flight and are destroyed. So American scientists built a rocket which can be used again and again. It is called the Space Shuttle.

The Space Shuttle has wings like an aircraft. However, it is launched like a rocket. Three powerful engines inside the Shuttle give it thrust, helped at first by two **booster rockets** fixed to the sides. When the Shuttle returns to Earth, the engines shoot gas forwards. This slows the Shuttle down. The Shuttle's wings help it to glide down through the Earth's atmosphere. The Shuttle lands on a very long runway.

▲ Most parts of the Space Shuttle can be re-used. The booster rockets drop off and float safely back to Earth held by parachutes. The fuel tank also drops off. This is the only part of the Shuttle which cannot be re-used.

▼ When the Shuttle comes in to land, it looks like an airliner.

43

Did you know?

Wind power

Hurricanes can blow at speeds of over 300 kilometres an hour. There is a huge amount of energy in hurricanes. If we could find a way to turn this energy into electricity, we would have all the power we need. A single hurricane in a day could supply a country like Canada with all the electricity it uses in a year!

The largest modern windmill

The world's largest modern windmill is at Howard's Knob, in North Carolina, in the United States. It has blades 60 metres long. It can generate enough electricity to supply about 2000 people.

Harrier jump jets

The British Harrier jet fighter flies like a plane but can take off without a runway. It uses a jet engine pointing downwards to push it up off the ground. This type of aircraft is called a VTOL, which stands for Vertical Take Off and Landing.

Hanging on

Pilots have kept hang gliders up in the air for over 24 hours at a time. Some have flown their gliders to a height of over 3000 metres.

An early helicopter

Leonardo da Vinci is the Italian artist who painted the *Mona Lisa*. He lived hundreds of years before people learned how to fly. Yet he designed and drew many flying machines, including a helicopter. He called it *helix pteron*, the Greek words for 'screw ' and 'wing'.

46

Deep-sea diving

People going underwater need special equipment to breathe. Bells filled with air and lowered into the water have been used since the 1500s. Diving bells are still in use today. Divers can move much more easily if they carry their air on their backs. Using this method and wearing special suits divers have reached depths of over 300 metres.

Wind 'farms'

We use fuels to make most of our electricity. Huge wind 'farms' can generate electricity cheaply. However, at the moment, we need about 1000 windmills to make as much electricity as one coal-burning **power station**. A thousand propellers whirling around would be very noisy! So they will need to be away from people.

Gases in the air

Air is made up of 78 per cent nitrogen and 21 per cent of oxygen. The remaining one per cent is made up of carbon dioxide, argon and other gases. There is usually some water vapour and often particles of sand and dust. Burning fuels adds unwanted gases and particles to the air. This pollution can be harmful to people, plants and animals. Over 51 per cent of the pollution comes from vehicles.

Sailing ships

Today, most of our ships are powered by fuels. These are expensive and we are using them up quickly. In the future, we may have little fuel left. The power of the wind is free and will never run out. So we may start to use sailing ships again to carry people and goods over the water. The sailing ships of the future will be controlled by computers, not by sailors.

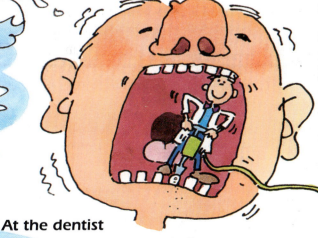

At the dentist

A dentist's drill works in a similar way to a windmill. Compressed air is used to drive a tiny turbine inside. This drives the drill shaft. In some types, the turbine blades float on a cushion of air.

Glossary

aerofoil: a shape, such as an aircraft's wing. It is rounded at the front and tapers to a point at the back

aileron: a movable flap on the wing of an aircraft. By moving the ailerons, the pilot can make the aircraft tilt over to one side

air conditioning: a way of keeping the air inside a building at the same temperature and wetness all year round

atmosphere: the mixture of gases, including air, which surrounds the Earth

atmospheric pressure: the pressing of the atmosphere on the Earth's surface

automatically: on its own

bank: tilt over to one side. An aircraft banks as it turns

battery: a device which stores electricity. Some batteries can be carried round easily

blades: the moving parts of a fan, propeller or turbine. Blades are often made of metal, and have sharp edges

booster rocket: an extra rocket engine used to increase or boost the speed of a space vehicle

breathing apparatus: containers of air or oxygen attached by a tube to a mask. People use breathing apparatus in places where there is little or no oxygen, such as in a burning building or high up a mountain

bronchi: the two tubes which lead from the windpipe to the lungs

carbon dioxide: a gas which humans breathe out and plants breathe in

central heating: a method of heating a building from a central place, such as a boiler room. Heat is carried around the building in pipes containing water or air

centrifugal: pushing out from the middle. A centrifugal fan pushes air outwards from the middle of the fan

compressed: squeezed into a small space. When compressed, air has more power

condense: change from a vapour into a liquid

cylinder: a hollow or solid tube shape

dew: drops of water which cover the ground in the early morning. Dew is formed from water vapour in the air which has cooled and condensed back to a liquid

diaphragm: the part of the stomach which moves to make us breathe

drag: the pull of the air on anything moving through the air

eardrum: the part of the ear which picks up sound

electricity: a form of energy which is used by many machines

electric motor: a machine which turns the energy of electricity into motion

elevator: a movable flap on the tail of an aircraft. Moving the elevators makes the aircraft climb or dive

energy: the ability to do work. We get our energy from food

evaporate: change from a liquid into a vapour. Heat makes water evaporate into the air

exhaust fumes: the gases and dirt which are given off by a motor vehicle

expand: get bigger. A liquid expands when it changes into a gas

filter: an air filter removes the dust and dirt from air. Some air filters are made of paper or cloth. They trap the dust and dirt while the clean air passes through

fuels: substances which are burnt to give heat and other forms of energy

gas: a substance which will move to fill any space. Air is made up of gases. Air fills all the space around us

gas burners: devices which burn fuel gas to give heat

gear wheel: a wheel with teeth or cogs around the edge. The cogs of each gear wheel fit between the cogs of other gear wheels. As each gear wheel turns it moves the other gear wheels as well

generator: a machine which makes electricity

gravity: the pull of the Earth. Gravity makes objects fall and gives them weight

grindstone: a heavy stone wheel which is turned to grind corn or sharpen knives

hang glider: a glider in which the pilot hangs from the wings

heart: the part of the body which pumps blood around the body

heating element: wires which become hot when electricity flows through them. A heating element turns the energy of electricity into heat

hover: hang in the air without moving

hovercraft: a vehicle or machine which is lifted from the ground by a cushion of air. A hovercraft can skim over land or sea without touching the surface

humidity: the amount of water vapour in the air

hydrogen: a very light gas which can be easily set on fire. Hydrogen joins with oxygen to form water

inner tube: a rubber tube filled with air which fits inside the outer cover of a tyre

lift: the upward push of the air on the wings of an aircraft

lungs: the parts of the body which are filled with air when we breathe in. This is where oxygen is taken into the blood

nitrogen: the main gas in air

nozzle: a small tube. When a liquid is pushed through a narrow nozzle, the liquid comes out as a spray

oxygen: the gas in the air which we need to live

parachute: an umbrella-shaped device which lets people and objects fall safely from great heights

pilot parachute: a small parachute which opens up quickly, pulling out the main parachute

piston: a closely fitting rod or disc which moves up and down inside a cylinder

plies: layers of cord which strengthen the rubber coating of tyre

pneumatic drill: a machine which uses compressed air to hammer a blade into the ground

pneumatic tyre: a tyre which contains compressed air

power point: a place where an electrical appliance connects to the electric power supply

power station: a building where electricity is made

propeller: a device with large blades which is turned around rapidly by an engine. Propellers push ships through the water and push hovercraft and aircraft through the air

pump: a device which sucks in a substance at one end and pushes it out at the other end

respiration: breathing

respiratory system: the parts of the body which we use when we breathe

right angle: the space between joining sides of a square. When you stand upright your body is at right angles to the ground

rotors: the long rotating wings of a helicopter

rudder: a flat piece of wood or metal fixed to the back of a ship, hovercraft or aircraft. The rudder can be turned to the right or left, and turns the vehicle

scuba: the letters stand for self-contained underwater breathing apparatus. It is used by divers to breathe underwater

shaft: a rod which carries the power from a motor, engine or windmill to a machine or vehicle. A mine shaft is a deep hole from the surface down to the mine

siren: a machine in which air is pushed through the holes in a rapidly rotating disc or drum. It makes a wailing noise

suction: the push which moves substances to fill a vacuum

supersonic: faster then the speed of sound. This means a speed of at least 1000 kilometres an hour

symbiosis: two groups of animals or plants living together and depending on each other

syringe: a simple pump which is used to give injections

tacking: sailing a ship on a zig-zag course so that the ship can move against the wind

temperature: how hot or cold a substance is. The temperature is usually measured in °C (degrees Celsius)

thermal: a warm stream of air which moves upwards

thermostat: a device which controls temperature, such as in an air conditioning or heating system

thrust: the push needed to move an aircraft or rocket through the air or space

tread: the deep pattern cut into the surface of a rubber tyre. The tread helps the tyre to grip the road

tubeless tyre: a tyre without an inner tube

turbine: a machine with blades which are spun rapidly by a moving gas or a liquid

turbofan: a powerful jet engine which has a huge fan turned by a turbine. The fan speeds up the flow of gases and so increases the thrust of the engine

vacuum: an empty space with no air

valve: a device which can open or close to control the flow of a liquid or gas

vent: an opening

ventilation: replacing old used air with fresh air

vibration: rapid movement backwards and forwards, such as that of a rubber band when twanged

Index